from Ida Leahy
Australia

A colourful ephemeral flora dominated by 'paper-daisies' in Central Australia. Extensive areas of arid inland Australia are transformed in this way following heavy rains.

An attractive flowering display of Epacris, Pultenaea *and* Leptospermum *in sclerophyllous vegetation at the entrance of Mount Richmond National Park near Portland, Victoria.*

AUSTRALIA'S NATIVE FLOWERS

Calothamnus
CLAWFLOWER
An attractive shrub with showy, red flowers belonging to the family Myrtaceae. This genus is entirely restricted to Western Australia.

Photography by Ken Stepnell

Text by Teresa James

NATIONAL
BOOK DISTRIBUTORS AND PUBLISHERS

Ken Stepnell has had a consuming interest in Australia's flora and fauna since he was given a Box Brownie camera at an early age. His interest grew and he has developed his hobby into a full-time career with a huge photographic library being used in many publications.

Ken travels more than 100 000 kilometres annually in search of new subjects and is continually fascinated by the variety of topics available in Australia.

Teresa James, born in England, obtained a degree with Combined Honours in Biology and Geography at Exeter University.

She came to Australia in 1978 and was immediately fascinated by the individualism of Australia's native flowers. She says they are more showy than the European or British varieties.

Teresa spends her leisure hours bushwalking and painting still-life. She is currently Acting Identifications Officer of the National Herbarium at the Royal Botanic Gardens.

A field of native flowers near Alice Springs in Central Australia.

Published by
National Book Distributors and Publishers
3/2 Aquatic Drive, Frenchs Forest, NSW 2086
First edition 1986
Reprinted 1987, 1988, 1989, 1991, 1994, 1996
Photography © Ken Stepnell 1986
Text © Teresa James 1986
Typesetting processed by Deblaere Typesetting Pty Ltd
Printed in Singapore by Kyodo Printing Co (S'pore) Pte Ltd

All rights reserved. Subject to the Copyright Act 1968, no part of this publication may be reproduced, stored in a retrieval system, or transmitted, in any form or by any means, electronic, mechanical, photocopying, recording or otherwise, without the prior written permission of the publisher.

National Library of Australia
Cataloguing-in-Publication data

Stepnell, Kenneth, 1931- .
    Australia's native flowers.

    Includes index.
    ISBN 1 86436 246 4

    1. Wild flowers—Australia. 2. Wild flowers—Australia—Pictorial works. I. James, Teresa. II. Title.

582.130994

AUSTRALIA'S UNIQUE FLORA

Australia is an island continent of about 7.7 million square kilometres which has been isolated for millions of years. The climate varies markedly from the tropical north to the temperate south and from the arid centre to the moist eastern coast. Such climatic differences, combined with changes to topography and soils, provide a wide range of environments for plants to inhabit, from tropical rainforests to barren deserts, alpine tracts and sandy plains. It is not surprising, therefore, that Australia has one of the most diverse and remarkable floras in the world.

There are some 15 000 species of flowering plants, 500 species of ferns, conifers and cycads and a further 12 500 species of non-vascular plants including mosses, algae, lichens and fungi. The native flowers of Australia, from the time of their first discovery, have been renowned for their uniqueness. The European botanists were fascinated by the strangeness of the plants taken back by earlier explorers. The originality of many Australian plants is well illustrated by the bizarre 'kangaroo paws' of Western Australia with their furry paw-shaped flowers, the spectacular 'Sturt's desert pea' and the colourful, unusual 'spider-flowers' of the wide-spread genus *Grevillea*.

Such individuality is reflected in the large number of endemic species, plants that are only found in Australia. It has been estimated that some 33 per cent of Australian genera and a remarkable 85 per cent of species are endemic. The peak of their diversity and endemism is found in south-west Western Australia where many examples abound of large, spectacular plant groups which are confined to this region. The genus *Dryandra*, for example, is a member of the family Proteaceae and contains some 60 different species, none found outside the region. The genus *Calothamnus* with 24 species is similarly confined to this remarkable area. These are shrubs belonging to the family Myrtaceae and are closely related to the well-known 'bottlebrushes'. In the same family are the 'feather-flowers' of the genus *Verticordia* which are very attractive shrubs with unique, feathery blossoms. There are some 50 species of *Verticordia* all confined to the south-west except for one.

Australia clearly has a unique and remarkable collection of wildflowers but how did such a flora develop on an island continent? To answer this question we must go back some 200 million years to the time when Australia was part of a large land mass called 'Gondwanaland' with South America, Africa, India, Antarctica and New Zealand. Much of Australia at this time was subtropical rainforest in a warm, moist climate. This original vegetation was presumably derived from immigration over a number of fronts within this land mass. About 110 million years ago Gondwanaland began to separate until only Australia, Antarctica and South America remained in contact. Australia then began to drift

Anigozanthos rufus
RED KANGAROO PAW
A distinctive and well-known plant of W.A. The unusual paw-shaped flowers are covered in deep red woolly hairs.

away some 45 million years ago in a northward direction resulting in a gradual change from southern plants of temperate climates to those adapted to arid and wet tropical climates. Under conditions of increased geographical isolation, decreased rainfall and greater seasonality, the original vegetation began to change although many plants of this ancient flora have survived to the present day in temperate refuges in south-east Australia, Tasmania, New Zealand and New Caledonia.

The continued northerly movement of Australia eventually brought the continent into contact with south-east Asia about 10 million years ago. A land bridge allowed migration of plants in both directions but particularly into Australia, having an important effect on the composition of the flora of the Northern Territory and Queensland. The evolution of the flora was also affected by geological and climatic processes within Australia such as glaciation during the last two million years. As a result, sea levels were lowered and further land bridges between Australia and the Indo-Malaysian region were formed. During periods of glacial and inter-glacial activity both temperature and rainfall would have fluctuated markedly resulting in the expansion and contraction of the Gondwanaland rainforest accordingly. The contraction of the rainforest would have led to a corresponding spread and dominance of drier vegetation types.

A showy display of annual plants dominated by Parakeelya 'paper-daisies' in inland W.A.

The Australian flora of the present day, therefore, is an amalgamation of three main elements. There is the relict Gondwanaland flora which has links back with New Zealand, Antarctica and South America, a more recent derived 'Australian' flora which has seen the unique development of many endemic plants and an 'Asian' element particularly seen in the northern part of the continent as a result of links with Asia. The relict Gondwanaland flora is best represented today in Tasmania. The island's geographical position, its isolation and the relatively mountainous terrain have retained an environment suitable for its preservation. A number of Gondwanaland plant groups such as *Nothofagus* 'southern beech' and *Coprosma* are represented in the Tasmanian flora. In contrast, the flora of the arid region of mainland Australia is an example of part of the more recent flora which has developed from the Gondwanaland stock since the onset of extensive arid conditions. The arid zone covers more than one-third of the continent including most of the interior and western part. This area is relatively flat and has less than 250 mm of rain a year. It is sparsely vegetated with grasses, annuals, tough shrubs and stunted trees. A spectacular ephemeral flora of annual plants dominates the desert vegetation after heavy rains and provides a mass of colour. Extensive areas are carpeted with such plants as the 'paper-daisies' *Helichrysum* species, *Calandrinia* species or 'parakeelya' and the well-known *Clianthus formosus*, 'Sturt's desert pea'.

Most people associate Australia with the two main plant groups which dominate the Australian landscape, the eucalypts and wattles. These are both plants of the more recently derived 'Australian' flora. There are some 600 species of *Eucalyptus* found in Australia, occurring mostly in open forest and woodland. Relatively few species are found in the arid regions except in favourable sites such as along streams or in isolated mountain ranges. Similarly few eucalypts have been successful along rainforest margins or in alpine habitats but elsewhere they have become adapted to a wide range of soils, topography and climate. They can range from tall single-stemmed forest trees some 60 m in height to multi-stemmed 'mallee' forms usually less than 10 m high.

Eucalyptus macrocarpa
ROSE OF THE WEST
A spectacular red or pink flowering eucalypt of the W.A. sandheaths.

Eucalyptus papuana
GHOST GUM
A distinctive eucalypt with striking white, smooth bark. This species is widely distributed throughout northern Australia.

Acacia baileyana
COOTAMUNDRA WATTLE
A bluish-grey shrub or small tree with bipinnate leaves and bright yellow flower heads. Originally occurring naturally in the Cootamundra district, but now widely cultivated and naturalised in many places.

The bark of the different species varies greatly, some species shed the dead bark annually leaving a completely smooth surface, whereas in others, the dead bark is persistent for many years. Several bark types can be recognised including gums, stringy-barks, bloodwoods, boxes, peppermints and iron-barks. The leaves are typically tough and leathery. The development of the leaves involves seedling, juvenile, intermediate and adult phases which can be recognised by differences in shape, size and colour. The flowers are quite distinctive with no petals and masses of showy stamens. A cap or 'operculum' covers the stamens in bud and later falls off to allow the stamens to protrude. The flowers are often white or cream in colour and inconspicuous, but striking red, pink, yellow and orange flowers do occur especially in the Western Australian species. The characteristic smell of eucalyptus oil is familiar to most people. This oil has been long used medicinally in Australia particularly in the treatment of colds but it has also been used successfully as an insect repellant and insecticide.

The wattles belong to the genus *Acacia*, the largest in Australia with over 800 species. They range from tall trees to small shrubs and occur in most vegetation types but are particularly dominant in arid and semi-arid regions where they tend to replace the eucalypts as the dominant woodland genus. The small attractive flowers are massed into fluffy globular heads or spikes and vary in colour from light yellow to deep orange. The normal bipinnate feathery leaves are replaced by 'phyllodes' in many of the Australian species. During evolution the leaf blade has been lost and the leaf stalk has become broader forming a phyllode, probably in response to increasing arid conditions. The fruit of the wattle is a pod which contains numerous hard seeds. These seeds were used extensively by the Aboriginals for food because they have a high protein content.

Other major plant groups of the more recent, derived 'Australian' flora include the following families: Myrtaceae, Proteaceae, Rutaceae, Epacridaceae, Fabaceae, Casuarinaceae and Xanthorrhoeaceae. Some of the more important genera of these families are outlined.

Banksia prionotes
ACORN BANKSIA (*see page 19*)

Banksia repens
CREEPING BANKSIA
An unusual shrub with horizontal branches at ground level and erect leaves and flower spikes. A widespread species in open sand heaths in W.A.

Myrtaceae

In addition to the eucalypts there are many other important genera in this family. The 'tea-trees' or *Leptospermum* species are a widespread and familiar group of shrubs with small leaves and masses of white or pink, five-petalled flowers. As the common name suggests the leaves were used to produce a 'tea' beverage in the times of the early settlers.

The 'paper-barks' belong to the genus *Melaleuca*, a widespread group of trees and shrubs in Australia. The bark consists of many papery-thin layers which can be easily stripped. The Aboriginals used the bark for many purposes including making rafts and canoes, to cover their shelters and to wrap food for cooking over the camp fire. Today, the various shades of bark are used to create pictures and wall decorations. The leaves are usually small with numerous oil dots, a feature typical of this family. The flowers consist of masses of conspicuous stamens arranged in heads of spikes similar to the closely related 'bottlebrushes'. Other important genera include *Beaufortia*, *Calothamnus*, *Chamelaucium*, *Callistemon* and *Verticordia*.

Proteaceae

This is a very important Australian family containing many of the more spectacular plants. There are some 750 species in 40 genera in the family, the vast majority being confined to Australia. Some of the more well-known genera include *Grevillea*, *Hakea*, *Banksia*, *Dryandra* and *Telopea*.

The 'spider-flowers' or *Grevillea* species are a fascinating group with over 250 species recognised. They have become very popular garden plants in recent years. The unusual flowers have a conspicuous pistil (female part of the flower) which is exserted from the tubular perianth or base. These flowers are rich in nectar and prove an irresistible attraction to birds, especially honey-eaters. The 'silky oak' is a large tree to 30 metres high growing in rainforest country of N.S.W. and southern Queensland but many are smaller shrubs growing in heaths and forests.

The *Hakeas* are very similar to the 'spider-flowers' with comparable leaves and flowers. They differ mainly in having fruits that are very woody and nut-like. Hakeas are mainly shrubs but some grow in inland areas as small trees called 'corkwoods'.

A very well-known group is the genus *Banksia* with over 60 species found only in Australia. They were named in honour of Sir Joseph Banks who accompanied Captain Cook while exploring the eastern coast of Australia. They vary from prostrate shrubs to large trees. The flower spikes are a mass of individual flowers with conspicuous pistils and no petals, in colours of yellow to orange, red, brown and purple.

The genus *Dryandra* is very closely related to Banksia and is endemic to Western Australia. They are a spectacular group of shrubs with interesting, toothed foliage and showy flower heads surrounded by bracts.

The State Floral Emblem of N.S.W. is the 'waratah', *Telopea speciosissima*, a spectacular wildflower with striking red flower heads. The scientific name *Telopea* means 'seen from afar' and refers to the spectacle it makes in the bush during spring. There are also three other species occurring in the eastern and southern States.

Rutaceae

A family containing many aromatic wildflowers such as the Boronia. The 'sweet-scented boronia', *Boronia megastigma*, is a very popular plant admired for its sweet smell and rich brown and gold flowers. It grows in swampy forest country in south-west Western Australia and is now grown commercially for the cut-flower trade. There are some 90 species of Boronia in Australia. They are mostly small shrubs with small, aromatic leaves and starry four-petalled flowers in shades from pale pink to red, brown, yellow and blue.

The 'waxflowers' or *Eriostemon* species are similar to the Boronias but differ in having waxy, five-petalled flowers in white, mauve or pink. The most spectacular species is *Eriostemon australasius*, 'pink waxflower', which has very pink flowers and grows on the east coast of Australia.

The flowers of the 'wild fuchsias' or *Correa* species are very different from the preceding genera. They are attractive bell-shaped flowers, often drooping, in shades of red, pink, green and yellow. The plants are bushy shrubs with hairy leaves and grow mainly in the eastern States.

Epacridaceae

This familiar 'heath' family contains over 330 native species in 28 different genera. They are typically small shrubs with densely packed, pointed leaves, tubular flowers and often berry-like fruit. The flowers vary mainly from white to pink or red and are very attractive. The genus *Epacris* has some 40 species including the 'common heath', *Epacris impressa*, which has long-lasting, striking pink or white flowers and is the official Floral Emblem of Victoria. The greatest development of the heath family is in the southern states and Tasmania. A few members of the family can grow to the size of a small tree such as *Richea* species in Tasmania and a species of *Dracophyllum* which grows on Lord Howe Island and reaches an incredible 20 metres in height.

Other important genera in the group are *Leucopogon* and *Styphelia*.

Fabaceae

The familiar 'pea' flowers of this family are closely related to the wattles. They are a very large and widespread family well represented in Australia by 1100 species in 136 genera. They can be trees, shrubs, herbs or climbing plants with either compound or single leaves or the leaves may be modified to phyllodes, spines or scales. The flowers are typically 'pea-like' with a large, upstanding petal or 'standard', two smaller lateral petals or 'wings' and two lower petals joined to form a keel. The fruit is a legume or pod coming in a great variety of forms. The famous 'Sturt's desert pea' is a member of this family as are also the genera of *Pultenaea*, *Dillwynia* and *Bossiaea* with their familiar 'egg and bacon' flowers which produce such a spectacular sight in bushland especially in spring.

Casuarinaceae

The distinctive conifer-like 'she-oaks' are medium-sized shrubs to small trees which occur in a variety of habitats from coastal moist forest to arid regions. They belong to the genera *Casuarina* and *Allocasuarina*. The name Casuarina refers to the resemblance of the long, drooping branchlets of these plants to the feathers of a cassowary. The leaves are typically reduced to tiny teeth on jointed, needle-like branchlets. The seeds are produced in conspicuous woody cones.

Clianthus formosus
STURT'S DESERT PEA (*see page 25*)

Stirling Range in W.A. dominated by the distinctive grass-tree Kingia australis, *Black Gin.*

Xanthorrhoeaceae

A familiar sight in heathland and sclerophyll forests in Australia are the 'blackboys', *Xanthorrhoea* species and the closely related *Kingia* of Western Australia. These conspicuous grasstrees have above ground trunks which are crowned with numerous thin leaves. The xanthorrhoeas have spear-like flowering spikes that may persist for long periods of time. The flowers are lily-like, creamy in colour and are produced in a mass along the spike.

A striking aspect of the majority of the plant families outlined so far is the prevalence of relatively small, rigid or hard leaves and small plant size. Such plants are described as being 'sclerophyllous'. This is an important feature of the more recent 'Australian' flora. It was originally believed that this was an adaptation to increasing arid conditions but more recent studies show that it is probably due to the poor, infertile soils. The Australian land surface has undergone weathering continuously for a long time, therefore conditions favouring sclerophyllous vegetation have become increasingly more widespread. This is well illustrated in the flora of the arid region where many of the plants are essentially sclerophyllous unlike in other desert areas of the world, where adaptations such as deciduousness, spinescence and succulence are more common. A unique form of sclerophyllout grass, 'spinifex' or 'hummock grass', has even evolved in the Australian desert areas. This situation can be directly related to the poor, siliceous soils so widespread across inland Australia. In contrast, where richer alkaline and saline soils occur the plant forms are closer to those found in the deserts of the rest of the world. The poor soil conditions peculiar to the Australian continent, therefore, have given the Australian flora a unique character.

Over much of Australia these sclerophyllous plants grow together in forest, woodland, scrub and heath communities. The forests are usually dominated by *Eucalyptus* species with tall, well developed trunks and flattish crowns. The canopy is almost continuous and the ground is well covered by sclerophyllous shrubs and herbs.

Two main types of forest can be distinguished. Dry sclerophyll forests are open forests with two main layers, a dominant tree canopy and a ground cover of drought-resistant shrubs and herbs. Wet sclerophyll forests are much taller with an intermediate layer of tall shrubs and small trees above the ground cover. Although the species present in the latter are drought-resistant they are not capable of surviving long dry periods.

Fire has always been a dominant feature of sclerophyll forests and as a result many extraordinary strategies have evolved by the plants in order to survive. There are two main adaptive strategies evident. Some plants such as the eucalypts are fire tolerant with a range of protective features. The bark is often highly heat-resistant especially the smooth-barked gums, and this insulation protects the 'epicormic' buds which underlay the bark. These buds remain dormant until canopy loss occurs through fire and then become active allowing the sprouting of new branches. If complete destruction occurs survival must be from underground stems and roots called 'lignotubers' which are present as conspicuous swellings at the base of the trunk. These have many dormant buds which are stimulated to grow by the destruction of parts of the plant above ground. Other species of eucalypts and sclerophyllous trees and shrubs such as banksia, hakea and casuarina are fire sensitive but produce large quantities of seed which is protected during the fire in the soil, or in woody fruits, and germinates afterwards. Many wattles and shrubs belonging to the 'pea' family are killed readily by the fire yet these are often the first plants to reappear. This is due to the large quantities of hard seeds they produce which lie dormant in the soil and litter for long periods until they are cracked open by the heat of the fire enabling the seed to absorb water and germinate. Some plants actually need fire to complete their life cycles, for example, the 'blackboys', 'Christmas bells' and many native orchids require fire to induce flowering.

In woodland communities the trees do not form a continuous canopy as in sclerophyll forests and the trees usually have rounded crowns and short trunks. The ground cover may consist of shrubs, herbs and grasses in varying frequencies.

Heathland communities are shrub dominated and rarely exceed a height of more than two metres. They are floristically rich, producing attractive flowering displays for much of the year. The sclerophyllous plants are very tough and wiry with numerous drought-resistant features. Herbs such as orchids and lilies are frequently found in these heaths.

There are many other important plant groups which are part of the remarkable Australian flora; some of the more interesting are the orchids, the trigger-plants and the insectivorous plants.

Orchids

The orchids are a very popular and interesting group of plants. There are some 700 species native to Australia including some large genera which are completely or almost exclusively Australian such as *Prasophyllum* ('leek orchids'), *Pterostylis* ('greenhoods') and *Caladenia* ('spider orchids'). Many of the orchids had their origins in the southern States and have gradually spread through the continent although the tropical northern areas are dominated by orchids with an Asian origin. There are two major groups of orchids, the terrestrial or ground orchids and the epiphytic orchids. The terrestrial orchids are the best represented in Australia. They typically die down every year and regrow the next, from an underground tuber. Epiphytic orchids are supported on the trunks or branches of trees or on rocks and are particularly common in the forests and rainforests of northern N.S.W. and Queensland. These orchids absorb water and minerals from the surface of the bark and take in oxygen directly from the air. The flowers of orchids are closely related to the lilies in having three sepals outside the three petals but, unlike the latter in which all six segments are generally similar in shape, the upper petal of the orchid is greatly modified and given the name of labellum (little lip). The labellum is often attractively coloured and various types of hairs and glands may be present. Orchids also differ from the lilies in having the style, stigma and stamen of the flower fused together to form a slender structure or 'column'.

Australian native orchids are protected plants in all States as many rare species are becoming extinct as bushland gives way to expanding settlements. In recent years, however, two remarkable new species have been discovered by accident. These are the underground orchids *Rhizanthella* and *Cryptanthemis*. The life cycles of these plants are completed in an amazingly few days from breaking the ground to the dispersal of the seeds.

Trigger-Plants

These plants belong to the genus *Stylidium*, a delightful group of herbs with over 100 different species found throughout Australia. They usually have basal grass-like leaves and flowers of white, cream, yellow, pink or red. The common name refers to an interesting and unique mechanism which is used to ensure fertilisation of the flower. The flowers are tubular with five spreading lobes, one of which is quite inconspicuous. The male and female parts of the flower are united to form a 'column' which is very sensitive. The anthers ripen first and produce pollen and when an insect touches the base of the column it springs across dusting the visitor with pollen. As the flower matures the female stigma pushes its way up and when an insect triggers the column, pollen is picked up from the visitor and so fertilisation occurs.

Insectivorous plants

Australia is rich in species of insectivorous plants belonging to several small plant families. They form a specialist group which have developed skilful means of trapping their prey, a subject that has fascinated botanists for many years. These plants usually live in water or swampy habitats where the

Caladenia carnea
PINK FINGERS
A very widespread and variable species of ground orchid with pale to deep pink flowers.

Stylidium graminifolium
GRASS TRIGGER PLANT (*see page 56*)

essential elements for growth such as oxygen, nitrogen and phosphorous are in short supply. By digesting small animals the carnivorous plants obtain these elements directly.

The 'sundews' belong to the family *Droseraceae* and as the name suggests the plants are covered by sticky dew-like drops exuded by tentacles on the leaves. When an insect touches just one of these tentacles the others close over the victim too. Once the insect has been digested the tentacles are released and the trap is reset. The leaves designed for trapping grow either as rosettes at the base of the plant or sometimes along the stems. The flowers are white, pink or red and are delicate and attractive.

The 'bladderworts' of the genus *Utricularia* belong to the family *Lentibulariaceae* and are small plants which grow around the margins of freshwater swamps and in bogs. The leaves often grow below the surface of the water and develop small bladders which trap and digest insects in the water. The bladders operate by pumping out water creating pressure outside against a side door. The touch of an insect springs a trap, the door opens and water rushes in carrying the victim with it and closing the door behind. The flowers of these plants are generally quite attractive with a large lower lip in colours of white, yellow, blue and purple. There are some 40 species found mainly in the tropical regions of Australia.

The 'pitcher-plants' of the families *Cephalotaceae* and *Nepenthaceae* trap their victims in 'jug-like' structures which are developed from the leaves. The insects are attracted to the traps by liquid in the pitcher and nectary glands around the rim. Once the victims fall into the pitcher they are soaked with the liquid inside and escape is almost impossible. The 'Western Australian pitcher plant', *Cephalotus follicularis*, has a very restricted occurrence on the extreme south-west coast of Australia where it grows in peaty swamps. The pitchers are squat and bristly and when in a sunny position change from green to a deep red colour. The 'tropical pitcher plant', *Nepenthes*, is the largest of all Australian carnivorous plants. It is a climber of the damp rainforest of north-east Queensland where it grows over other plants by means of tendrils, the tips of which develop into the pitchers. The largest pitchers are only 20 cm long but are known to have trapped small birds and mammals. This plant is the only representative in Australia of a large genus of 70 known species.

Australia has inherited a unique collection of native flowers. The people of Australia are obligated, therefore, to preserve and safeguard this valuable resource. Although there has been growing community awareness of conservation needs in recent years, a much greater involvement is still required.

Man has been influencing the vegetation of Australia over many years. Aboriginal man has lived in Australia for at least 40 000 years during which time he has affected the vegetation by the use of fire and to some extent by the cultivation of food plants. The most destructive influence, however, has been experienced since European colonisation in 1788. Since this time the uncontrolled spread of agriculture, grazing, urbanisation and industrial development, forestry, mining, roadworks, use of herbicides, horticultural collecting and competition from introduced plants has had a devastating effect on the vegetation. It is estimated that only 30% of the original vegetation present in 1788 remains untouched today. In less than 200 years some 78 species of plants are believed to have become extinct and 201 are currently endangered. Alien plants introduced from other countries into Australia are estimated to make up 10% of our present flora. These aliens have been introduced by man intentionally and accidentally.

Problems arise when these introduced plants escape from cultivation areas to compete with native species. They establish most readily on disturbed sites such as road verges and wasteland, however examples such as *Lantana camara* (lantana), *Ulex europaeus* (gorse), *Opuntia* spp. (prickly pear) and *Ligustrum* spp. (privet) illustrate how they can become established in relatively undisturbed bushland. Another example is *Chrysanthemoides monilifera* (boneseed) which was introduced into Sydney from South Africa as an ornamental and later used in dune stabilisation and revegetation after beach mining. Its great capacity for invading natural bushland was recognised too late, eradication programmes were commenced but still today it is very much a problem in coastal areas of N.S.W.

In recent years a number of conservation measures have been taken but these mainly involve the preservation of certain plant communities at particular localities, often where the scenery is very spectacular. There should be a greater emphasis on the conservation of whole floristic types such as the temperate and sub-tropical rainforests of eastern Australia. These rainforests are remnants of the ancient Gondwanaland flora and contain some of the most primitive flowering plants in the world and are the surviving remains of the stocks from which most of the modern Australian flora has been derived. Surely it should be considered a matter of national pride that what remains of them should be conserved.

Public education and awareness is of utmost importance in the campaign for preserving the Australian flora. There have been films, television documentaries, journals and books released on the native flora and fauna in recent years and posters, calendars, stationery and printed textiles promoting the subject are being produced commercially. The Botanic Gardens in Australia have an important role to play promoting the cultivation of native plants and in growing those species known to be endangered. Interest in the cultivation of Australian plants by the home gardener has been tremendous in recent years but this has not always been the case. Before 1950 there were very few nurseries selling a range of native plants but from about 1960 onwards, probably due largely to the establishment of the Society for Growing Native Plants, there began to be a great upsurge of interest. Severe drought conditions in the late 1960s proved the value of growing native plants as many gardens with introduced plants suffered badly. This led to a dramatic increase in the demand for native plants and the number of nurseries selling Australian plants correspondingly grew. This interest has lasted during the 1970s and early 1980s and it is now commonplace to see natives grown in urban areas and widely planted in parks and public gardens and as street trees. The cut-flower trade has also placed more emphasis on natives in recent years. Areas are being planted with species of *Dryandra*, *Banksia*, *Agonis*, and *Telopea*, to name a few. The market for these unique flowers is increasing both overseas and in Australia.

If the present high levels of interest and concern about the Australian flora can continue, it must help to preserve and protect the unique collection of wildflowers in Australia for future generations.

Acacia acinacea
GOLD-DUST WATTLE
The 'wattles' are a very diverse and widespread group of plants with over 800 species in Australia. The leaves are typically bipinnate and feathery in appearance or reduced to phyllodes (modified leaf stalks) which vary greatly in form from small to needle-like to large, flat and leaf-like. The individual fluffy yellow to orange flowers are massed into attractive globular heads or spikes. This species is a bushy shrub with long arching branches and masses of bright yellow flower heads produced in spring and early summer. It grows in woodland or open scrub in parts of N.S.W., Vic. and S.A.

Acacia aneura
MULGA
A bushy shrub or small tree often forming dense, extensive stands in the arid regions of most States. The phyllodes are typically narrow and bluish-green and the bright yellow flowers are arranged in spikes, appearing after good rains. 'Mulga' is a useful plant, the foliage being used as fodder for livestock, the wood for fencing and in the past for making Aboriginal implements and weapons. The seeds were also used as a source of food by some Aboriginals.

Acacia pycnantha
GOLDEN WATTLE
One of the most attractive of the 'wattles' with masses of fragrant golden flower heads. A large shrub or small tree native to a few restricted areas in Australia but now naturalised in many parts of the eastern States. The bark is one of the richest sources of tannin in the world.

Acacia verticillata
PRICKLY MOSES
A prickly, dull-green erect or trailing shrub with whorled phyllodes and dense flowering spikes. It is usually found in damp situations in forest or woodland in the southern States.

Anigozanthos humilis
CAT'S PAWS
The 'kangaroo' or 'cat's paws' of southwest W.A. are lily-like perennial herbs with striking paw-shaped furry flowers in colours ranging from yellow to orange, green and red. This species is a low growing plant of the sand plains with orange and yellow flowers arranged in a single spike.

Anigozanthos manglesii
RED AND GREEN KANGAROO PAW ▲
The best known of the 'kangaroo paws' and the W.A. Floral Emblem. The striking contrast of the brilliant green of the flowers and the red at the base and along the stems makes it a most distinctive plant. Common on the sandy coastal plains and growing in profusion in bushland around Perth, flowering from winter to early summer.

Anigozanthos viridis
GREEN KANGAROO PAW ◄
As the name suggests this species has flower heads and stems clothed in various shades of green ranging from lemon-yellow to deep emerald. It inhabits damp clay depressions in a restricted area on the coastal plains of W.A.

Astroloma pinifolium
PINE HEATH ►
A dense bush with fine pine-like leaves and lime-yellow or reddish, green-tipped tubular flowers produced along the stems from spring to autumn. The flowers are followed by succulent fruits which are sweet when ripe and provide a source of food for the Aboriginals. Widespread in heaths and forests of N.S.W., Vic. and Tas.

Banksia baueri
POSSUM BANKSIA
The 'banksias' are among some of Australia's most fascinating plants with their showy, cylindrical spikes of densely packed, spirally arranged flowers. There are over 60 species, all but one occurring only in Australia. This unusual species has large, furry flower spikes which nestle among the foliage resembling small furry animals. The flowering spikes are the largest of all the banksia species, up to 40 cm long, and vary in colour from lemon-yellow and mauvish-grey to a bright tan. It is found mainly in open sandy heaths in southwest W.A.

Banksia coccinea
SCARLET BANKSIA
A slender shrub or small tree with spectacular scarlet and grey, squat flower heads appearing from winter to summer. Common on exposed coastal sand plains or in moist gullies inland in W.A. This species is used extensively in the cut-flower trade.

Banksia prionotes
ACORN BANKSIA
A beautiful tree growing in the sandy soils of southwest W.A. The bright orange styles of the flowers are released from the bottom of the spike upwards, giving an acorn-like appearance. The leaves have interesting saw-toothed margins.

Banksia serrata
SAW BANKSIA
This species, discovered by Sir Joseph Banks at Botany Bay in 1770, forms a gnarled spectacular tree or large shrub in sandy heath or forest in near-coastal districts of N.S.W., Vic. and Tas. It has flat leaves with saw-toothed margins and large grey to yellow flower spikes.

Bossiaea cinerea
SHOWY BOSSIAEA ▲
An interesting genus of some 45 species belonging to the large 'pea' family. The leaves are extremely variable and sometimes absent but the flowers are characteristically 'pea'-like in shades of yellow and red. This species with its showy flowers in spring is found in near-coastal heaths of the eastern States.

Beaufortia decussata
BEAUFORTIA ▲
The 'beaufortias' are related to the well-known 'melaleucas' or 'paper-barks' and occur naturally in W.A. but are now widely cultivated throughout Australia. They have small crowded leaves and brush-like spikes or heads of red, pink or mauve flowers. This species has deep red flowering spikes appearing mainly during the summer months.

Caladenia aphylla
LEAFLESS ORCHID ▼
Caladenias are attractive ground orchids and most of the 80 known species are found only in Australia. They vary widely in form, size and flower colour. This species is the first to flower in the new season during March and April. As the name suggests this species does not produce the usual single leaf characteristic of the genus. It grows in 'she-oak' forests in the southwest of W.A.

Caladenia dilatata
GREEN COMB SPIDER ORCHID ▲
One of the best known 'spider orchids', this widespread species is found in heathlands and open forests in most States. The flowers are solitary, green and white with a maroon-tipped labellum and a distinguishing green comb on either side.

Caladenia flava ◀
COWSLIP ORCHID
The only bright yellow *Caladenia*. A widespread species in southwest W.A. The flowers are 3-5 cm across with variable red markings, and are produced during September and October.

Caladenia patersonii var. *longicauda* ▶
WHITE SPIDER ORCHID
A graceful 'Spider Orchid' with large flowers 10-20 cm across, often sweetly scented. A common and widespread species in open bushland in southwest W.A.

Calandrinia balonensis
BROAD-LEAF PARAKEELYA
A small succulent herb with showy purple flowers and broad fleshy leaves. It is common on sand plains and sand dunes in Central Australia where it frequently makes attractive massed displays over extensive areas.

Calandrinia polyandra
PARAKEELYA
This species is similar to the 'Broad-Leaf Parakeelya' but differing in the more succulent and narrower leaves. The flowers open for one day only during spring. It is widespread in the drier areas of W.A. and S.A.

Caleana major
LARGE DUCK ORCHID
A slender plant to 30 cm high with large green or dull reddish-brown flowers resembling a duck in flight. This species forms small colonies in dry open forests in most States.

Calectasia cyanea
BLUE TINSEL LILY
An unusual member of the 'lily' family with a shrubby habit and crowded narrow leaves. The flowers are quite spectacular with six shiny, blue-purple petals contrasting with the erect, golden anthers. It grows in heathland and open forest in Vic., S.A. and W.A.

Callistemon speciosus
ALBANY BOTTLEBRUSH
One of the well-known and widely cultivated 'bottlebrushes' with the typical dense flower spikes. This is an erect shrub with red flowers produced in spring and early summer, growing on the margins of swamps and watercourses mainly in the Albany district of W.A.

Calochilus robertsonii
PURPLISH BEARD ORCHID
The 'beard orchids' are easily distinguished by the hairy or bearded labellum of the flowers. This species is a robust plant with up to 15 small yellow-green flowers with reddish-brown markings, produced in spring. It occurs in most States.

Celmisia sp.
SILVER SNOW DAISY
The 'snow daisies' of Australia are a distinct feature of the alpine regions where they occur in a variety of habitats. This form has silvery foliage and masses of white flowers produced in summer.

Calothamnus sp.
CALOTHAMNUS
A hardy group of bushy plants related to the 'bottlebrushes' with flowers in various shades of red or cream growing along the stems, usually to one side. The 24 recognised species all occur only in W.A.

Chiloglottis gunnii
COMMON LARGE BIRD ORCHID
There are seven species of *Chiloglottis* in Australia, all of which have a pair of broad, basal leaves and a solitary green to purplish flower resembling a young bird's wide-open mouth. This species is found in moist subalpine forests in Tas., Vic., and N.S.W.

Chorizema sp.
FLAME PEA
A genus of some 20 species occurring mainly in southwest W.A. They are shrubs and twining plants with masses of attractive red, yellow, orange and pink 'pea' flowers.

Clianthus formosus
STURT'S DESERT PEA
One of the most spectacular of all Australian plants with its vivid red petals and glossy, swollen black or deep purple centre. The individual 'pea' flowers are up to 10 cm long and hang in clusters at the top of short, erect stems. A plant of the arid inland areas of most States where it may be locally common especially after rains in late winter and spring. It is the Floral Emblem of S.A.

Cochlospermum sp.
KAPOK
A small group of tall shrubs and small trees found only in the tropical parts of Qld. N.T. and W.A. In drier monsoonal regions these plants often lose their leaves during the dry season. The large yellow, five-petalled flowers are very showy and are often produced when the leaves are absent.

Conospermum mitchellii
VICTORIAN SMOKEBUSH
The 'smokebushes' are an interesting group of plants with some 50 species found in all States, but the majority occurring in W.A. They have tubular flowers growing in crowded spikes at the ends of the branches, in colours of white, cream, yellow and grey to blue. This species is an impressive shrub with profuse creamy-white flowers growing in sandy lowland heaths and at higher altitudes in Vic. and S.A.

Comesperma spinosum
SPINY MILKWORT
A genus of some 24 species endemic to Australia consisting of shrubs and twining plants with pink to blue flowers with large wing-like sepals. This species is a thorny shrub occurring in W.A.

Correa reflexa
COMMON CORREA or NATIVE FUSCHIA
A very widespread and variable shrub ranging widely through the eastern States from areas of mallee scrub to mountain forests. The flowers are large and bell-shaped in many shades of red with green tips or uniformly red, pink or yellow-green appearing in spring.

Corybas dilatatus
VEINED HELMET ORCHID
This most unusual miniature orchid has a single heart-shaped leaf and a solitary hooded flower 2-3 cm across on a short stalk. It favours cool mountain gullies and shaded coastal areas in the southern States. The flowers appear in late winter and spring. There are 10 species of 'helmet orchids' in Australia with representatives in most States.

Corybas unguiculatus
SMALL HELMET ORCHID
Similar to the 'Veined Helmet Orchid' but with smaller reddish-purple flowers about 1 cm across, bird-like in appearance and on a long stalk. It inhabits humus in shaded, near-coastal sites.

Craspedia sp.
BILLY BUTTONS
Members of the diverse daisy family, this group of plants are herbs with mainly basal leaves and yellow flowers without petals, arranged in dense, globular heads on long stalks. They usually occur in small colonies and grow during the cooler months and become conspicuous in spring when the flower heads develop. There are five species currently recognised and representatives are found in all States.

Dampiera sp.
DAMPIERA
This group of plants are herbs with flowers in many shades of blue, closely related to *Leschenaultia*. The flowers are five-petalled and fan-shaped and, although small, are produced in great numbers giving very attractive displays. They are found in a variety of habitats in most States although the majority are found in W.A.

Daviesia brevifolia
LEAFLESS BITTER-PEA
A member of a large genus of some 70 species all endemic to Australia. This species is a broom-like heathland bush without conspicuous leaves and profuse small 'pea' flowers, delightfully coloured apricot to wine red. The flowers are followed by triangular pods. It is found in Vic., S.A. and W.A.

Daviesia corymbosa
CLUSTERED BITTER-PEA
A shrub to 2 m high with variable leaves and attractive clusters of bright yellow and red flowers. A widespread species on sandstone in forests in N.S.W.

Dichopogon strictus
CHOCOLATE LILY ▶

A small plant with long, fleshy leaves and small, white to mauve flowers along a slender stem. The flowers have a delicious chocolate-like or caramel scent. It is found in temperate parts of most States in open grassland, woodland and open forest. A small genus with only two species, both endemic.

Dillwynia sericea
SHOWY PARROT-PEA

The 'parrot-peas' are heath-like shrubs with small, narrow leaves and yellow 'pea' flowers with orange and red markings. There are about 26 species known, occurring in most States but particularly in Victoria. This species has showy apricot-yellow flowers in leafy spikes and is widespread in the southern States. ▶

Diuris longifolia
COMMON DONKEY ORCHID or WALLFLOWER DIURIS

This group of ground orchids have attractive flowers ranging from white to yellow or orange with red, purple, or brown markings. The flowers have two upper petals which extend like ears giving rise to the common name 'donkey orchids'. This species has large yellow and purple flowers occurring in a wide variety of soils and conditions in most States. ◀

Diuris sulphurea
SULPHUR DONKEY ORCHID

This species has sulphur yellow flowers with brown markings. A very variable spring-flowering plant in most States. ▼

Drakaea sp.
HAMMER ORCHID
This plant belongs to a group of orchids found only in W.A. with several unnamed species. The flowers are visually insignificant due to their subdued colouration but nevertheless they are highly specialised for insect pollination. The labellum of the flower resembles a wingless female wasp at the top of a grass stem, thus attracting the male wasp. All species have small heart-shaped leaves which lie flat on the ground.

Dodonaea microzyga
BRILLIANT HOPBUSH ▼
There are over 50 species of 'hopbushes', most of which are endemic to Australia. The flowers are small and inconspicuous but are always followed by winged capsules which are often attractively coloured. This species is a showy shrub with brown to red coloured fruit massed along the branches and grows in Central Australia.

Drosera menziesii
PINK RAINBOW or SUNDEW ▼
The 'sundews' are an interesting group of insectivorous plants which trap small insects on sticky hairs around the edges of the leaves and then digest them. There are about 70 different species all endemic to Australia with the majority occurring in southwest W.A. where they grow in moist, boggy sites. The flowers can be showy as in this species with pink to reddish flowers up to 2.5 cm across.

Dryandra falcata
DRYANDRA
The 'dryandras' are a distinctive group of plants with some 60 species found only in southwest W.A. The leaves are toothed in many attractive patterns and the flowers of yellow, orange and red grow in heads which are surrounded by a ring of bracts. This species is a shrub with holly-like leaves and yellow flower heads 4-5 cm across produced in spring. It grows in sand or gravel soils in heathland.

Dryandra formosa
SHOWY DRYANDRA
A well-known bushy shrub or small tree with bright yellow-orange flower heads to 10 cm across and long, narrow leaves toothed to the midrib. This is one of the most spectacular wildflowers of the Stirling Ranges in W.A. and is commonly cultivated throughout Australia.

Elythranthera brunonis
PURPLE ENAMEL ORCHID
A spectacular orchid with glossy, rich purple flowers produced in spring which tend to fade after fertilisation until almost pink. A widespread and common species in southwest W.A.

Epacris impressa
COMMON HEATH
The Australian heaths are typically low growing to erect shrubs, with small, crowded, sharply pointed leaves and white, pink or red tubular flowers. These popular plants are long-blooming from winter through to summer and are common and widespread in southeast Australia. The pink form is the Victorian Floral Emblem.

Epacris longifolia
NATIVE FUCHSIA ▲
A straggling shrub with attractive red tubular flowers with whitish tips. It is confined to the sandstone areas of N.S.W., common around Sydney, growing on ledges and cliffs particularly in moist sites.

Eremophila latrobei
CRIMSON TURKEY BUSH ◄
The 'emu' or 'turkey bushes' are shrubs or small trees with attractive, bell-shaped flowers in many colours varying from white to pink, red, orange, yellow, blue and purple. They are found throughout the drier parts of Australia as they are drought-tolerant. This species is a densely leafy shrub with red to purple flowers produced mainly in spring and autumn.

Eriochilus cucullatus
PARSON'S BANDS ►
A dainty ground orchid distinguished by 2 white, lateral petals which resemble the pendant strips of neckband formerly worn by the clergy. The flowers are usually solitary and have a delicate perfume. This is one of four species endemic to Australia. It is a widespread species in the eastern States, particularly abundant in Vic.

Eucalyptus ficifolia
W.A. RED FLOWERING GUM ▲
This is one of the best known ornamental eucalypts with spectacular flowers produced in summer. The flowers are of various shades of red to white and are produced so profusely that the foliage is almost hidden. A tree 8-14 m high restricted in the wild to a small area on sandy soil near Albany, W.A.

◄
Eucalyptus forrestiana
FUCHSIA GUM
A small tree or mallee with smooth grey bark which is shed in long strips. This species is distinguished by large, four-sided winged buds and fruit which are bright red at first turning brownish with time. They appear together in profusion on pendulous stalks in summer and autumn. A very ornamental tree restricted to a small area in the wild in W.A.

Eucalyptus megacornuta
WARTY YATE
A slender, stiffish tree 7-12 m high with unusual warty, gherkin-like buds produced on strap-like peduncles. Large, yellow-green flowers appear in profusion in winter and spring. This form of the species grows only on a few lateritic hillsides of the Ravensthorpe Range, W.A.

Glossodia major
LARGE WAX-LIP ORCHID ▲
A widespread ground orchid of southeast Australia producing brilliant springtime displays. The flowers are mostly blue but sometimes deep magenta or white and are produced at the top of the flowering stem. A solitary, narrow leaf at ground level arises annually from a small underground tuber.

Gastrodia sesamoides
POTATO ORCHID ▲
A slender, leafless plant rising from a potato-like tuber which was used by the Aboriginals as a food source. The flowers are bell-like with a swollen base, cinnamon brown in colour, with a whitish tip. There are only two species of 'potato orchids' in Australia, both occurring in cool woodlands and moist montane forests.

Glischrocaryon sp.
GOLDEN PENNANTS ▶
A small group of showy herbs with numerous slender stems and clusters of loose, bright yellow flowers which wave in the slightest breeze. There are four species present in Australia growing mainly in the dry, sandy soils of the inland areas.

Gompholobium ecostatum
DWARF WEDGE PEA
Gompholobium species produce some of the largest of the 'pea' flowers in Australia. This species is a low-growing shrub with beautiful apricot to orange-red flowers appearing during spring and summer. It is found growing in heathland in Vic. and S.A.

Gossypium australe
DESERT ROSE
A pubescent shrub with white to pink flowers with a dark reddish centre and reddish anthers. An attractive plant growing in gravelly soils or on sandy plains in grassland or woodland in W.A., N.T. and Qld.

Gossypium sturtianum
STURT'S DESERT ROSE
The Floral Emblem of the N.T., this ornamental shrub has large, pale purple hibiscus-like flowers appearing during winter and spring and rounded leaves dotted with tiny oil glands. It is highly drought resistant growing in stony ground in Central Australia.

Grevillea alpina
MOUNTAIN GREVILLEA
The 'grevilleas' are amongst the most attractive of Australian plants and include over 250 different species, most of which are endemic to Australia. They vary greatly from low-growing trailing shrubs to tall trees. The flowers may be almost any colour except blue and grow in many different arrangements including the typical 'spider-like' type. This species grows at higher altitudes in N.S.W. and Vic. The large, showy, orange-red flowers are typical of the 'Grampians form' of this species.

Grevillea dimorpha
FLAME GREVILLEA
A shrub with olive-like leaves, silvery below, and with brilliant red 'spider-flowers' appearing in autumn to late spring. This species is found only in the Grampian Ranges of Vic.

Grevillea juncifolia
HONEYSUCKLE SPIDER-FLOWER
A small tree with large, bright orange, velvety flower spikes produced in spring. It is found as isolated plants in arid areas of Australia.

Gunniopsis kochii
GUNNIOPSIS ▲
A low-growing succulent, annual herb occurring in rocky soil in central eastern S.A. The flowers have four petals which are an attractive pale pink on the inside.

Hakea sp.
HAKEA ▼
The genus *Hakea* is very large, consisting of some 130 species all endemic to Australia and distributed in every State. They vary from small shrubs to large trees and have flowers similar to *Grevillea*. This is a tree species of inland Australia with long cylindrical leaves and yellow flowers in showy clusters.

Hakea teretifolia
DAGGER HAKEA
This species is a shrub with stiff, pungent leaves, profuse clusters of hairy, whitish flowers and unusual narrow, sharply pointed woody fruits. It is a widespread plant occurring in most States. ▼

Helichrysum spp.
PAPER-DAISIES or EVERLASTINGS
This genus comprises the largest group of the 'paper-daisies' with over 100 different species. They are annuals or perennial shrubs with either large flower heads surrounded by papery white, yellow, pink or brown petal-like bracts or numerous small flower heads clustered into a larger head. A widespread group of plants present in all States. The attractive colours of the flowers and their long-lasting qualities have made them very popular in dried flower arrangements.

Helichrysum davenportii
STICKY EVERLASTING
This is one of the larger flowered species with the white to pink flower heads up to 3 cm across borne singly on leafless stems. The leaves are basal, up to 5 cm long and aromatic. It is found in the drier areas of W.A., S.A. and N.T. where it flowers after heavy rains.

Hibbertia fasciculata
BUNDLED GUINEA-FLOWER ▲
The brilliant yellow and orange blooms of the 'guinea-flowers' provide a striking display in the bush. The flowers have five spreading petals and can be up to 7 cm across in some species. This species with its clusters of leaves and abundant sessile flowers grows in sandy heaths in the southern States.

▶
Hibbertia glaberrima
DESERT GUINEA-FLOWER
A small shrub with narrow, glossy leaves and deep yellow flowers produced in spring. It is found in the rocky ranges of Central Australia.

◀
Hibiscus panduriformis
YELLOW HIBISCUS
There are some 50 species of native *Hibiscus* which vary from small trees to low growing shrubs. The typical hibiscus-like flowers range in colour from white to yellow, orange, pink and red. This species is a soft hairy shrub with beautiful deep yellow petals and a red centre. It is found along river banks in W.A. and also in northern Australia.

Hovea sp.
HOVEA
A group of plants belonging to the large and variable 'pea' family. There are 14 species of *Hovea*, all endemic to Australia. They are erect shrubs with attractive blue to purple 'pea' flowers borne close to the stem and found in all States.

Hybanthus floribundus
SHRUB VIOLET
A variable small shrub with abundant pale blue, lilac or whitish flowers belonging to the violet family. The flowers are five-petalled but four are inconspicuous while the lowest petal is enlarged, streaked with violet and yellowish at the base. A widespread shrub in the southern States.

Isopogan latifolius
DRUMSTICK
The 'drumsticks' are an interesting group of plants with over 30 species occurring in most States but the majority are found in W.A. The common name refers to the round seed cone which is borne at the end of a long stem and resembles a drumstick. This species is a tall shrub of the rocky Stirling Ranges in W.A. The flowers are magnificent bright pink to purple and up to 8 cm across, appearing in spring.

Kennedia prostrata
RUNNING POSTMAN
The genus *Kennedia* contains over 13 species of trailing perennial herbs with trifoliate leaves and 'pea' flowers in colours of red, purple to black and yellow. This species has bright red flowers appearing in spring and summer and is a common plant of coastal districts and inland, in moist situations in temperate Australia.

Kingia australis
BLACK GIN
A type of 'grass-tree' found only in W.A., closely related to the 'blackboys'. A slow-growing plant with a trunk to 5 m tall with long stiff leaves and small flowers growing in dense globular heads at the end of a long stalk. Widespread in parts of W.A.

Lambertia formosa
MOUNTAIN DEVIL or HONEY FLOWER
There are 10 species of *Lambertia* in Australia, all of which are erect shrubs with sharply pointed leaves and clusters of yellow, orange or red tubular flowers. Only one species is found outside of W.A.

It is commonly known as the 'mountain devil' referring to its woody fruit which has a short beak and a long horn on either side. The flowers are reddish in groups of about seven, enclosed in long red bracts. A long-flowering shrub restricted to the sandy soils of the N.S.W. coast and the Blue Mountains.

Lambertia inermis
CHITTICK
The 'chittick' is a shrub to 6 m tall with heads of cream or orange flowers found growing in deep sands in heath in W.A. ▶

Lechenaultia biloba
BLUE LECHENAULTIA
This is the best known of the *Lechenaultias*, a chiefly W.A. group of some 19 species with characteristic tubular flowers with five spreading lobes in a variety of colours. This species is a shrub with profuse, delicate flowers of various shades of blue which make massed displays during spring in southern and central W.A. ◀

Lechenaultia formosa
RED LECHENAULTIA
A small plant with a wide range of colour forms from red to orange, yellow and pink. A common plant in parts of W.A. ▼

Leptosema chambersii
THE UPSIDE-DOWN PLANT ▲
A small group of plants with larger than usual 'pea' flowers in various shades of red. This species is a leafless shrub with spiny, terete branches and long red flowers which rest on the soil around the base of the plant. The Aboriginals used the abundant nectar produced by the flowers as a food source. A widespread plant on sandy soils in Central Australia.

Leptospermum nitidum
SHINY TEA-TREE
An attractive shrub belonging to the well-known 'tea-tree' group. This species has large white, five-petalled flowers, 2-3 cm across, appearing in late spring. A frequent plant in Tas. and on parts of the Grampians in Vic. ▼

Lyperanthus serratus
RATTLE BEAKS
A ground orchid arising annually from an underground bulb. Although the plants are tall they are often inconspicuous because of their dull-coloured flowers and their habit of growing up through dense bushes. This is one of four species found in Australia, growing in the heavier rainfall areas of southwest W.A.

Lysiana exocarpi
HARLEQUIN MISTLETOE
A parasitic plant growing along the stems of a wide range of host plants including the 'wattles' and the 'she-oaks'. The flowers are tubular with red, waxy petals united for over half their length and five or six yellowish-green spreading lobes at the tip. The flowers appear in summer or late winter and are followed by succulent berries ripening from green to red or black. ◄

Lysinema ciliatum
CURRY FLOWER
A small, slender shrub in the heath family with small leaves and white, tubular, curry-scented flowers enclosed for most of their length by brown sepals and bracts. This is one of only five species, all confined to southwest W.A. ▼

Macropidia fuliginosa
BLACK KANGAROO PAW ▲
This striking plant is closely related to the other 'kangaroo paws' in the genus *Anigozanthos*. It has strap-like leaves and tall flowering stems with unusual black-felted, greenish-yellow flower clusters. It is found growing on the sand plains in heath and woodlands between Perth and Geraldton, W.A.

▶

Maireana triptera
THREE-WINGED BLUEBUSH
A member of a large group of plants of inland arid Australia commonly known as 'bluebushes' because of their bluish-green appearance. This species is a small shrub with fleshy leaves and attractive glossy, papery, green, orange or red fruits densely arranged along the branches turning dark brown to black when dry.

◀
Melaleuca sp.
PAPER-BARK
This well-known group of plants contains some 140 species throughout Australia. The flowers, in which the stamens are the most conspicuous part, resemble the 'bottlebrush' inflorescence of the genus *Callistemon*. The colour range is great with various shades of white, cream, yellow to orange, red, pink and mauve. The bark of the trees or shrubs is typically papery and spongy. A very widely cultivated group.

Microseris lanceolata
NATIVE YAM or NATIVE DANDELION ▲
A herb belonging to the daisy family which rises annually from a fleshy root and produces solitary large yellow flower heads at the ends of long stalks. The fleshy roots have a sweet coconut flavour and were eaten by the Aboriginals. A widespread plant throughout temperate Australia.

Myriocephalus stuartii
POACHED EGG DAISY ◀
An attractive annual daisy with greyish foliage and yellow flower heads surrounded by white papery bracts. This is one of 12 species endemic to Australia which provide spectacular flowering displays after heavy rains over extensive areas in the dry inland.

Orthoceras strictum
HORNED ORCHID ▶
A ground orchid with several grass-like leaves and greenish-brown flowers in spikes. The upright sepals of the flowers resemble horns giving rise to the common name 'horned orchid'. A widespread species growing in damp, often swampy conditions throughout the eastern States.

Oxylobium alpestre
ALPINE OXYLOBIUM
The 'oxylobiums' are a group of 'pea' flowers which produce showy displays of yellow to orange-red flowers in spring and early summer. There are some 25 species with representatives in most States. This species is found at higher altitudes often among rocks in Vic. and N.S.W.

Patersonia longiscapa
LONG PURPLE FLAGS
A native iris with delicate and striking blue-purple flowers borne on long stems in early summer. It grows in damp heaths in Vic., S.A. and Tas.
▲

Pimelea rosea
ROSE BANJINE
There are over 80 species of *Pimelea* found throughout Australia, all being attractive shrubs with colourful clusters of small tubular flowers. The 'rose banjine' has soft, pink flower heads 2-2.5 cm across growing on the coastal plain from Perth to Albany in W.A. White-flowered forms are found away from the coast in forest areas.
▶

Platylobium obtusangulum
COMMON FLAT PEA
A member of the 'pea' family, this low spreading shrub has triangular, arrow-shaped leaves and bright red and yellow flowers which are followed by flat pods. Common in heaths and heathy understoreys especially near the coast in the southern States. ▼

Prasophyllum sp.
LEEK ORCHID
A large group of ground orchids characterised by a single, leek-like leaf and long racemes of small, inconspicuous flowers. When viewed through a magnifying glass these flowers are quite spectacular. There are about 100 species occurring in all States.

Prostanthera aspalathoides
SCARLET MINT BUSH
A strongly aromatic shrub unique among most of the Australian mint bushes in having bright red flowers. The flowers of this group are typically tubular with lobed petals and usually in shades of white, mauve or purple. This species has a scattered distribution in the drier, sandy areas of Vic., N.S.W. and Tas.

Pterostylis nutans
NODDING GREENHOOD
A member of a large group of ground orchids with leaves either in a basal rosette or stem-clasping and quaint, green flowers marked with red, brown or purplish colours. The unique feature of this group is the 'hooded' form of the flower. This species is one of the commonest found in most States, flowering during winter and spring.

Pterostylis vittata
BANDED GREENHOOD ORCHID
A very variable species often forming spectacular displays between May and September in coastal scrub. The flowers are green with red to brown bands or all reddish-brown. Found in Vic., S.A., W.A. and Tas.

Ptilotus exaltatus
SHOWY FOXTAILS or MULLA MULLA
A showy plant with purplish, fluffy flower spikes and soft hairy leaves and stems. Flowers appear in spring and early summer and during favourable years cover extensive areas of arid inland Australia. There are some 100 species widely distributed through the mainland.

Pultenaea gunnii
GOLDEN BUSH-PEA
The bright yellow, orange and red 'pea' flowers of *Pultenaea* species are a familiar sight in the bush. There are some 115 species, all endemic to Australia. They vary from low, mat-like herbs to large erect shrubs and can be found in most States. This species is aptly named from the profusion of rich orange-yellow flowers that appear in spring almost concealing its tiny rounded leaves. It is a sprawling shrub preferring damp situations in Vic., N.S.W. and Tas.

Radyera farragei
BUSH HIBISCUS ▶

A velvety shrub with large, round, cordate leaves and large flowers related to the popular *Hibiscus*. The flowers are pale purple and darker in the centre, the five petals overlapping. It is found as isolated plants or in small groups growing on sandy soils in arid areas.

Ricinocarpos gloria-medii
GLORY OF THE CENTRE

A spreading shrub with attractive white or creamy flowers, up to 2.5 cm across, borne in leafy clusters. The flowers are unisexual with both male and female flowers occurring on the same plant. The photograph above shows a male flower. A restricted species growing at the top of a steep, rocky scree slope at one locality at Simpsons Gap, Central Australia. ◀

Scaevola sp.
SCAEVOLA

A large group of attractive shrubs found chiefly in W.A. The flowers are typically five-petalled in a hand-like or fan arrangement in colours of white, blue, purple or pink. ▼

Schoenia cassiniana
SCHOENIA
A type of 'paper daisy' closely related to *Helichrysum*. An annual with glandular, aromatic leaves and flower heads 2-3 cm across with pink, papery bracts surrounding a yellow centre. The only species in this genus, it is found widely in W.A., S.A. and N.T.

Senecio gregorii
FLESHY GROUNDSEL
The 'groundsels' are members of the diverse 'daisy' family with some 40 species in Australia and thousands of other species distributed throughout the world. They vary from small herbs to bushy shrubs with typical daisy flowers in different shades of yellow. This species is a succulent, blue-green annual which becomes abundant after favourable winter rains, often dominating extensive areas of Central Australia.

Solanum sp.
SOLANUM
Plants belonging to the genus *Solanum* include not only the culinary potato and tomato but also many herbs and shrubs native to Australia. The flowers are typically blue-purple with conspicuous yellow stamens followed by tomato-like fruits. The stems and leaves are often covered with prickles.

Sphaerolobium vimineum
LEAFLESS GLOBE-PEA
A rush-like, leafless plant with attractive racemes of yellow or reddish 'pea' flowers followed by globular pods. It is widely distributed through the eastern States in heathland. The majority of the remaining 11 species are confined to W.A.

Stackhousia monogyna
CREAMY STACKHOUSIA
A herb with small, white tubular flowers arranged in erect candle-like spikes, fragrant at night. A widespread species in the eastern States. This is one of 22 species found throughout Australia.

Stylidium graminifolium
GRASS TRIGGER PLANT
The 'trigger plants' are an interesting group of plants with some 130 species in Australia, all remarkably designed for insect pollination. The flowers are tubular with five spreading lobes, one being inconspicuous. The 'column' of the flower is sensitive to touch and when triggered by an insect it springs across and secures the visitor who deposits or receives pollen. This species is a widespread perennial with tufted, grass-like leaves and numerous pink flowers appearing in a long spike during summer. It occurs in all States except W.A. on a variety of soil types.

Stylidium macranthum
LARGE-FLOWERED TRIGGER PLANT
A short-stemmed herb with large, rose-purple flowers on long stalks found in W.A.

Stypandra glauca
NODDING BLUE-LILY
A perennial herb belonging to the 'lily' family with bluish grass-like leaves and racemes of nodding, blue, star-like flowers with conspicuous yellow anthers. A widespread species on sandstone and poor soils in Qld. N.S.W., Vic. and S.A.

Styphelia behrii
FLAME HEATH
A small group of attractive heath plants with flat, pointed leaves and tubular flowers of pink, red, yellow, cream and green. As the flowers develop, the five lobes roll back exposing the stamens. This species has spectacular vivid, scarlet flowers followed by succulent fruits. It occurs in Vic., S.A. and N.S.W.

Styphelia pinifolia
PINE HEATH
A low bush with crowded, pine-like leaves and tubular yellow flowers tipped with green. It is found scattered through Vic., N.S.W. and Tas.

Synaphea sp.
SYNAPHEA
A small group of only six species confined to W.A. They are tufted shrubs with deeply lobed leaves and elongated yellow flower-spikes.

Telopea speciosissima
WARATAH
The Floral Emblem of N.S.W., this is a very striking plant when flowering in spring. The flowers are deep red and arranged in a dense, terminal head surrounded by red bracts. A widespread species in N.S.W. but never common and found on sandy soils.

Tetratheca sp.
BLACK-EYE SUSAN
An attractive showy shrub with boronia-like flowers, pink to purplish-red with a darker centre. An entirely Australian group with 39 species represented in all States except the N.T.

Thelymitra antennifera
RABBIT'S EARS or LEMON ORCHID
A member of a large group of orchids commonly called 'sun orchids' as the flowers open and close with the sun. There are some 40 species occurring in Australia. This species is conspicuous by the brown 'rabbit-ear' appendages of the column. It occurs in moist sites or near granite outcrops in drier areas in most States.

Thelymitra crinita
BLUE LADY ORCHID
A lovely plant with long spikes of sky-blue flowers up to 4 cm across which decorate the bush in late spring. A widespread species in forests and woodlands in W.A.

Thelymitra ixioides
DOTTED SUN ORCHID
The commonest of the 'sun orchids' occurring in all States. The flowers are variable in colour from bright blue to pink or occasionally white and the petals are often spotted.

Thysanotus multiflorus
MANY-FLOWERED FRINGED LILY
The genus *Thysanotus* belongs to the 'lily' family and contains 32 species of perennial herbs. The flowers are typically three-petalled, mauve to purple with each of the petals attractively fringed. This species is a tufted plant with showy clusters of mauve flowers found in W.A.

Tribulus cistoides
TRIBULUS
A low-growing herb with silky, divided leaves and solitary, five-petalled, yellow flowers up to 3.5 cm across. The flowers are followed by large spiny fruits. It is found growing in sandy, stony ground in W.A., N.T. and Qld.

Utricularia dichotoma
FAIRY APRONS
Small but attractive plants which grow in swamps and bogs and on wet rock ledges. The basal leaves develop small bladders which are designed to trap and digest small aquatic animals. The purple flowers have an enlarged lower lip and are arranged along slender stems appearing in spring and summer. This species is common in the southern States.

Verticordia sp.
VERTICORDIA
A spectacular group of plants with some 50 species found almost exclusively in W.A. They are bushy shrubs with small, five-petalled flowers which are delicately fringed. Flower colour varies from white to cream, orange, pink to red. The flowers are produced so profusely that the leaves are often completely obscured.

Xanthorrhoea sp.
GRASS TREE
The 'grass trees' or 'blackboys' are conspicuous plants of the Australian landscape. They are slow-growing plants with a trunk above or below the ground and long, narrow, tough leaves crowded at the tops of the stems. The flowers are white to cream on a long, thick flowering spike.

INDEX

Acorn Banksia (*Banksia prionotes*) 10, 19
Albany Bottlebrush (*Callistemon speciosus*) 23
Alpine Oxylobium (*Oxylobium alpestre*) 50

Banded Greenhood Orchid (*Pterostylis vittata*) 52
Banksia
  Acorn (*Banksia prionotes*) 10, 19
  Creeping (*Banksia repens*) 10
  Possum (*Banksia baueri*) 18
  Saw (*Banksia serrata*) 19
  Scarlet (*Banksia coccinea*) 18
Beaufortia (*Beaufortia decussata*) 20
Billy Buttons (*Craspedia* sp.) 27
Bitter-Pea
  Clustered (*Daviesia corymbosa*) 28
  Leafless (*Daviesia brevifolia*) 28
Blackboys (*Xanthorrhoea*) 12
Black-Eye Susan (*Tetratheca* sp.) 60
Black Gin (*Kingia australis*) 12, 44
Black Kangaroo Paw (*Macropidia fuliginosa*) 48
Bladderworts (*Utricularia* sp.) 14
Bluebush, Three-Winged (*Maireana triptera*) 48
Blue Lady Orchid (*Thelymitra crinita*) 60
Blue Lechenaultia (*Lechenaualtia biloba*) 45
Blue Tinsel Lily (*Calectasia cyanea*) 23
Boneseed (*Chrysanthemoides monilifera*) 14
Boronia, Sweet-Scented (*Boronia megastigma*) 11
Bottlebrush, Albany (*Callistemon speciosus*) 23
Brilliant Hopbush (*Dodonaea microzyga*) 30
Broad-Leaf Parakeelya (*Calandrinia balonensis*) 22
Bundled Guinea-Flower (*Hibbertia fasciculata*) 42
Bush Hibiscus (*Radyera farragei*) 53
Bush-Pea, Golden (*Pultenaea gunnii*) 52

Calothamnus (*Calothamnus* sp.) 7, 24
Casuarinaceae 11
Cat's Paws (*Anigozanthos humilis*) 16
Chittick (*Lambertia inermis*) 45
Chocolate Lily (*Dichopogon strictus*) 29
Clawflower 11
Clustered Bitter-Pea (*Daviesia corymbosa*) 28
Common Donkey Orchid (*Diuris longifolia*) 29
Common Flat Pea (*Platylobium obtusangulum*) 50
Common Heath (*Epacris impressa*) 11, 32
Common Large Bird Orchid (*Chiloglottis gunnii*) 24
Cootamundra Wattle (*Acacia baileyana*) 9
Correa, Common (*Correa reflexa*) 26
Cowslip Orchid (*Caladenia flava*) 21
Creamy Stackhousia (*Stackhousia monogyna*) 55
Creeping Banksia (*Banksia repens*) 10
Crimson Turkey Bush (*Eremophila latrobei*) 33
Curry Flower (*Lysinema ciliatum*) 47

Daisy, Poached Egg (*Myriocephalus stuartii*) 49
Dampiera (*Dampiera* sp.) 28
Dandelion, Native (*Microseris lanceolata*) 49
Desert Guinea-Flower (*Hibbertia glaberrima*) 42
Desert Rose (*Gossypium australe*) 36
  Sturt's (*Gossypium sturtianum*) 37
Donkey Orchid
  Common (*Diuris longifolia*) 29
  Sulphur (*Diuris sulphurea*) 29
Dotted Sun Orchid (*Thelymitra ixioides*) 60
Drumstick (*Isopogon latifolius*) 43
Dryandra 7, 10
Dryandra (*Dryandra falcata*) 31
  Showy (*Dryandra formosa*) 31
Dwarf Wedge Pea (*Gompholobium ecostatum*) 36

Epacridaceae 11
Eucalyptus 8
Everlastings (*Helichrysum* spp.) 41

Fabaceae 11
Fairy Aprons (*Utricularia dichotoma*) 61
Flame Grevillea (*Grevillea dimorpha*) 38
Flame Heath (*Styphelia behrii*) 57
Flame Pea (*Chorizema* sp.) 25
Flat Pea, Common (*Platylobium obtusangulum*) 50
Fleshy Groundsel (*Senecio gregorii*) 54
Foxtails, Showy (*Ptilotus exaltus*) 52
Fuchsia Gum (*Eucalyptus forrestiana*) 34
Fuchsia
  Native (*Correa reflexa*) 11, 26
  Native (*Epacris longifolia*) 33

Ghost Gum (*Eucalyptus papuana*) 9
Globe-Pea (*Sphaerolobium vimineum*) 55
Glory of the Centre (*Ricinocarpos gloria-medii*) 53
Gold-Dust Wattle (*Acacia acinacea*) 15
Golden Bush-Pea (*Pultenaea gunnii*) 52
Golden Pennants (*Glischrocaryon* sp.) 35
Golden Wattle (*Acacia pycnantha*) 16
Gorse (*Ulex europaeus*) 14
Grass Tree (*Xanthorrhoea* sp.) 62
Grass Trigger Plant (*Stylidium graminifolium*) 13, 56
Green Comb Spider Orchid (*Caladenia dilatata*) 21
Grevillea 10
  Flame (*Grevillea dimorpha*) 38
  Mountain (*Grevillea alpina*) 38
Groundsel, Fleshy (*Senecio gregorii*) 54
Guinea-Flower
  Bundled (*Hibbertia fasciculata*) 42
  Desert (*Hibbertia glaberrima*) 42
Gum
  Fuchsia (*Eucalyptus forrestiana*) 34
  W.A. Red Flowering (*Eucalyptus ficifolia*) 34
Gunniopsis (*Gunniopsis kochii*) 40

Hakea (*Hakea* sp.) 10, 40
  Dagger (*Hakea teretifolia*) 40
Hammer Orchid (*Drakaea* sp.) 30
Harlequin Mistletoe (*Lysiana exocarpi*) 47
Heath
  Common (*Epacris impressa*) 11, 32
  Flame (*Styphelia behrii*) 57
  Pine (*Styphelia pinifolia*) 58
Hibiscus
  Bush (*Radyera farragei*) 53
  Yellow (*Hibiscus panduriformis*) 42
Honey Flower (*Lambertia formosa*) 45
Honeysuckle Spider-Flower (*Grevillea juncifolia*) 38
Hopbush, Brilliant (*Dodonaea microzyga*) 30
Horned Orchid (*Orthoceras strictum*) 49
Hovea (*Hovea* sp.) 43

Insectivorous plants 13

Kangaroo Paw
  Black (*Macropidia fuliginosa*) 48
  Green (*Anigozanthos viridis*) 17
  Red (*Anigozanthos rufus*) 7
  Red and Green (*Anigozanthos manglesii*) 17
Kapok (*Cochlospermum* sp.) 26

Lantana (*Lantana camara*) 14
Large Duck Orchid (*Caleana major*) 22
Large-Flowered Trigger Plant (*Stylidium macranthum*) 56
Large Wax-Lip Orchid (*Glossodia major*) 35
Leafless Bitter-Pea (*Daviesia brevifolia*) 28
Leafless Globe-Pea (*Sphaerolobium vimineum*) 55
Leafless Orchid (*Caladenia aphylla*) 20
Lechenaultia
  Blue (*Lechenaultia biloba*) 45
  Red (*Lechenaultia formosa*) 45

Leek Orchid (*Prasophyllum* sp.) 51
Lemon Orchid (*Thelymitra antennifera*) 60
Lily
  Blue Tinsel (*Calectasia cyanea*) 23
  Chocolate (*Dichopogon strictus*) 29
  Many-Flowered Fringed (*Thysanotus multiflorus*) 61
Long Purple Flags (*Patersonia longiscapa*) 50

Many-Flowered Fringed Lily (*Thysanotus multiflorus*) 61
Mint Bush, Scarlet (*Prostanthera aspalathoides*) 51
Mistletoe, Harlequin (*Lysiana exocarpi*) 47
Mountain Devil (*Lambertia formosa*) 45
Mountain Grevillea (*Grevillea alpina*) 38
Mulga (*Acacia aneura*) 15
Mulla Mulla (*Ptilotus exaltus*) 52
Myrtaceae 10

Native Fuchsia
  (*Correa reflexa*) 26
  (*Epacris longifolia*) 33
Native Dandelion (*Microseris lanceolata*) 49
Native Yam (*Microseris lanceolata*) 49
Nodding Blue-Lily (*Stypandra glauca*) 57
Nodding Greenhood (*Pterostylis nutans*) 52

Orchid 13
  Banded Greenhood (*Pterostylis vittata*) 52
  Blue Lady (*Thelymitra crinita*) 60
  Common Donkey (*Diuris longifolia*) 29
  Common Large Bird (*Chiloglottis gunnii*) 24
  Cowslip (*Caladenia flava*) 21
  Dotted Sun (*Thelymitra ixioides*) 60
  Green Comb Spider (*Caladenia dilatata*) 21
  Hammer (*Drakaea* sp.) 30
  Helmet, Small (*Corybas aunguiculatus*) 27
  Helmet, Veined (*Corybas dilatatus*) 27
  Horned (*Orthoceras strictum*) 49
  Large Duck (*Caleana major*) 23
  Large Wax-Lip (*Glossodia major*) 35
  Leafless (*Caladenia aphylla*) 20
  Leek (*Prasophyllum* sp.) 51
  Lemon (*Thelymitra antennifera*) 60
  Nodding Greenhood (*Pterostylis nutans*) 52
  Parson's Bands (*Eriochilus cucullatus*) 33
  Pink Fingers (*Caladenia carnea*) 13
  Potato (*Gastrodia sesamoides*) 35
  Purple Enamel (*Elythranthera brunonis*) 32
  Purplish Beard (*Calochilus robertsonii*) 24
  Rabbit's Ears (*Thelymitra antennifera*) 60
  Sulphur Donkey (*Diuris sulphurea*) 29
  Underground (*Rhizanthella, Cryptanthemis*) 13
  White Spider (*Caladenia pattersoni* var. *longicauda*) 21
Oxylobium, Alpine (*Oxylobium alpestre*) 50

Paper-barks (*Melaleuca* sp.) 10, 48
Paper-Daisies (*Helichrysum* sp.) 8, 41
Parakeelya (*Calandrinia polyandra*) 8, 22
  Broad-leaf (*Calandrinia balonensis*) 22
Parrot-Pea, Showy (*Dillwynia sericea*) 29
Parson's Bands (*Eriochilus cucullatus*) 33
Pine Heath
  (*Astroloma pinifolium*) 17
  (*Styphelia pinifolia*) 58
Pink Fingers (*Caladenia carnea*) 13
Pink Rainbow (*Drosera menziesii*) 30
Pitcher-Plant
  Tropical (*Nepenthes*) 14
  W.A. (*Cephalotus follicularis*) 14
Poached Egg Daisy (*Myriocephalus stuartii*) 49
Possum Banksia (*Banksia baueri*) 18
Potato Orchid (*Gastrodia sesamoides*) 35
Prickly Moses (*Acacia verticillata*) 16
Prickly Pear (*Opuntia*) 14
Privet (*Ligustrum* spp.) 14
Proteaceae 10
Purple Enamel Orchid (*Elythranthera brunonis*) 32
Purplish Bear Orchid (*Calochilus robertsonii*) 24

Rabbit's Ears (*Thelymitra antennifera*) 60
Rattle Beaks (*Lyperanthus serratus*) 47
Red and Green Kangaroo Paw (*Anigozanthos manglesii*) 17
Red Kangaroo Paw (*Anigozanthos rufus*) 7
Red Lechenaultia (*Lechenaultia formosa*) 45
Rose Banjine (*Pimelea rosea*) 50
Rose of the West (*Eucalyptus macrocarpa*) 9
Running Postman (*Kennedia prostrata*) 44
Rutaceae 11

Saw Banksia (*Banksia serrata*) 19
Scaevola (*Scaevola* sp.) 53
Scarlet Banksia (*Banksia coccinea*) 18
Scarlet Mint Bush (*Prostanthera aspalathoides*) 51
Schoenia (*Schoenia cassiniana*) 54
Shrub Violet (*Hybanthus floribundus*) 43
Sclerophyll Forests 12
Shiny Tea-Tree (*Leptospermum nitidum*) 46
Showy Bossiaea (*Bossiaea cinera*) 20
Showy Dryandra (*Dryandra formosa*) 31
Showy Foxtails (*Ptilotus exaltus*) 51
Showy Parrot-Pea (*Dillwynia sericea*) 29
Silver Snow Daisy (*Celmisia* /sp.) 24
Small Helmet Orchid (*Corybas unguiculatus*) 27
Smokebush, Victorian (*Conospermum mitchellii*) 26
Snow Daisy, Silver, (*Celmisia* sp.) 24
Solanum (*Solanum* sp.) 54
Spider-Flowers (*Grevillea*) 10
Spiny Milkwort (*Comesperma spinosum*) 26
Stackhousia, Creamy (*Stackhousia monogyna*) 55
Sticky Everlasting (*Helichrysum davenportii*) 54
Sturt's Desert Pea (*Clianthus formosus*) 8, 11, 25
Sturt's Desert Rose (*Gossypium sturtianum*) 37
Sulphur Donkey Orchid (*Diuris sulphurea*) 29
Sundew (*Drosera menziesii*) 13, 30
Sweet-Scented Boronia (*Boronia megastigma*) 11
Synaphea (*Synaphea* sp.) 58

Tea-Tree, Shiny (*Leptospermum nitidum*) 46
Tea-Trees (*Leptospermum*) 10
Three-Winged Bluebush (*Maireana triptera*) 48
Tribulus (*Tribulus cistoides*) 61
Trigger Plants (*Stylidium*) 13
  Grass (*Stylidium graminifolium*) 13, 56
  Large-Flowered (*Stylidium macranthum*) 56
Tropical Pitcher Plant (*Nepenthes*) 14
Turkey Bush, Crimson (*Eremophila latrobei*) 33

Upside-Down Plant (*Leptosema chambersii*) 46

Veined Helmet Orchid (*Corybas dilatatus*) 27
Verticordia (*Verticordia* sp.) 7, 61
Victorian Smokebush (*Conospermum mitchellii*) 26
Violet, Shrub (*Hybanthus floribundus*) 43

Wallflower Diuris (*Diuris longifolia*) 29
Waratah (*Telopea speciosissima*) 11, 59
Warty Yate (*Eucalyptus megacornuta*) 34
Wattle (*Acacia* sp.) 8
  Cootamundra (*Acacia baileyana*) 9
  Gold-Dust (*Acacia acinacea*) 15
  Golden (*Acacia pycnantha*) 16
Waxflower, Pink (*Eriostemon australasius*) 11
W.A. Pitcher Plant (*Cephalotus follicularis*) 14
W.A. Red Flowering Gum (*Eucalyptus ficifolia*) 34
White Spider Orchid (*Caladenia patersonii* var. *longicauda*) 21

Xanthorrhoeacea 12

Yam, Native (*Microseris lanceolata*) 49
Yellow Hibiscus (*Hibiscus panduriformis*) 42